HUMAN EXISTENCE AND THE UNIVERSE

BY

KONG DERICK NJIKEH

DEDICATION

To Humanity

TABLE OF CONTENTS

INTRODUCTION

CHAPTER ONE: The Components of Human..........6

CHAPTER TWO: Communication between the Components of Human……………………………...12

CHAPTER THREE: Human and the Environment….17

CHAPTER FOUR: Creation and Evolution……..........22

CHAPTER FIVE: Relationship between the Brain and the Heart……………………………………………..27

CHAPTER SIX: Sleep and Dreams………….............31

CHAPTER SEVEN: The Mentally Offset Human and the Society…………………………………………...33

CHAPTER EIGHT: Electromagnetic Wave and Gaseous Communication between Individuals…............39

CHAPTER NINE: Origin of a Genius Ability………..49

CHAPTER TEN: A Genius and the Educational System……………………………………………....51

CHAPTER ELEVEN: A Genius and Socialization....54

CHAPTER TWELVE: The Universe………………57

CHAPTER THIRTEEN: Time Traveling………......60

CHAPTER FOURTEEN: Time Distortion…………63

CHAPTER FIFTEEN: Origin of Gravity and Weightlessness in Space……………………………66

CHAPTER SIXTEEN: Derician Trialism…………...69

CHAPTER SEVENTEEN: Procreation Evolution of Living Organisms on Earth……………………….77

CHAPTER EIGHTEEN: Consciousness……………79

CHAPTER NINETEEN: The Purpose of Human Existence……………………………………………83

CHAPTER TWENTY: Existence after Death……...85

ANNEX: Key Sentences………………………...88

INTRODUCTION

This book is an in-depth continuation of the book entitled "Human Existence". It talks about the synergistic nature of theology, science and philosophy, the purpose of human existence in this universe and some processes occurring in the universe.

Specifically, it brings out ideas on the origin of human and his composition, the evolution of human and the interaction of human with the environment.

Also, it talks about who a genius is and the origin of his ability, the interaction between a genius and the educational system and the process of socialization in a genius.

Furthermore, it talks about the origin of the universe, the concept of time, the process of time traveling and time distortion by human and the origin of gravity and weightlessness.

Finally, it talks about the trialism nature of human, the origin of consciousness, the purpose of human existence and the existence of human after death.

CHAPTER ONE

THE COMPONENTS OF HUMAN

The creator also known as super being, of human being created human into three components which coexist with each other and have a reversible system of communication between each other for human existence in this physical universe. These components are defined by human according to two fields of thought which are; the Theologian who is a human who gives invisible explanations through faith, of processes occurring in this physical universe and the Scientist who is a human who gives physical explanations through proof, of the processes occurring in this physical universe.

According to the Theologian, these components of human are called the Spirit, the Soul and the Body components.

The spirit component being the invisible and intangible component which is in direct relationship with the creator, acting in the invisible universe and through which everything that human performs in this visible universe arises from. It is therefore the main component, for it generates and coordinates the activities of the other components.

The next component, the soul, is the visible and intangible component which links the spirit and the body together. It conveys information from the spirit to the body and from the body to spirit. This means that the soul acts in the invisible and the visible universe. It is the component that separate from the body when human existence in action in this visible universe comes to an end.

The last component, the body, is the visible and tangible component which is responsible for all the actions carried out by human in this visible universe, and it marks the physical existence of human. There for the body generates no information but just receive commands from the spirit through the intermediary of the soul.

These components that make up human can be separated as human moves from the visible to the invisible universe of existence and it is carried out in stages.

Firstly, in this visible universe, the three components exist together which gives human a life to live where he is visible, tangible and can carry out actions.

Secondly, there comes a stage where the soul separates from the body, living only the spirit and the soul as the components that determine human existence while the body exists no more. In this stage, human is visible, intangible and cannot carry out any action in this visible universe meaning he has no life to live. This stage is also like a transition of human existence from the visible to the invisible universe.

Then thirdly, there comes a stage where the spirit separate from the soul, living only the spirit as the component that determines human existence while the soul exist no more. In this stage, human is invisible, intangible and cannot carry out any action in this visible universe meaning that he is now existing only in the invisible universe.

According to the Scientist, these components which make up human are called the conscious, the subconscious and the unconscious components.

The conscious component is also referred to as the mind which is the origin of all thoughts responsible for the generation of all actions carried out by human. The conscious component cannot be seen and cannot be touch.

The next component, the subconscious component, is an intermediary between the conscious and the unconscious components. I believe that it represents the impulses by which messages are transferred from the conscious to the unconscious component for action to take place. The subconscious component can be seen using instruments but it cannot be touch.

The last component, the unconscious component, is the one responsible for carrying out actions commanded by the conscious component through the intermediary of the subconscious component. It is made up of matter and it can be seen and touch. This is the component that gives human a life to live in this physical universe.

There is a reversible system of communication among these components which follows a definite linear pattern according to their arrangement, for human existence in this physical universe.

From the understanding of these different fields of thoughts that is, the Theologian and the Scientist, the three components that makes up human are the same but just have different names.

Therefore, I can conclude that the spirit component is also called the conscious component which is represented by the mind, the soul component is also called the subconscious component which is represented by impulses and the body component is also called the unconscious component which is represented as visible actions carried out.

Also, the Theologian believe that there is life after leaving this physical universe (life after death) and the Scientist don't believe that there is life after leaving this physical universe. But I can say that, there is no life after leaving this physical universe since life depends on the body component which is no more but instead, there is existence of human in the non-physical universe after leaving this physical universe. And also, there is existence of human in the non-physical universe before existing in this physical universe.

CHAPTER TWO

COMMUNICATION BETWEEN THE DIFFERENT COMPONENTS OF HUMAN

For human existence in this physical universe, the three components that make up human are arranged by the creator in a fixed defined pattern which is; the spirit, the soul and the body component. It is made that there is a reversible system of information communication between these components following this pattern of arrangement. That is to say, the spirit communicates with the soul and the soul communicates with the spirit, the soul then communicates with the body and the body communicates with the soul. This can be represented in a linear fashion as follows:

$$\text{Spirit} \longleftrightarrow \text{Soul} \longleftrightarrow \text{Body}$$

This implies that as human exist in this physical universe; there is no direct interaction between the spirit and the body and vice versa.

As human lives a life in this physical universe, there comes a time when the communication of information from the body to the soul is disrupted. This can be represented as:

Spirit ⟷ Soul ⟶ Body

I believe that this stage occurs during the period when there is a disease affection of matters which makes up the human body which can be cause by a variety of factors such as infectious, traumatic, tumoral, genetic, and other environmental factors. This disruption of communication from the body to the soul is reestablished by treating the underlining disease leading to a reformation of the body.

Also, as human lives a life in this physical universe, there comes a time when there is disruption of communication from the soul to the spirit. This can be represented as:

Spirit ⟶ Soul ⟷ Body

I believe that this stage occurs when there is affection of the soul by genetic, biological and environmental factors without affection of the body. This is seen in the "mentally offset" human who by himself do acknowledge that he has no disease.

To reestablish this disruption, the spirit hyper stimulates the soul which leads to the production of hallucination and delusion. The soul in turn hyper stimulates the body part associated which also leads to a probable increase in size of that body part.

I believe that, the normal human can help in the reestablishment of this communication from the soul to the spirit through Thoughts Transference (Telepathy) and language communication, all of which falls under psychotherapy and not through drugs and instrumental means which instead destroys the effort made by the spirit.

So, as it is seen from the above explanations, as human lives a life in this physical universe, the communication between the components that make up human can be disrupted in a definite single direction and it can also be reestablished. That is, the body from the soul or the soul from the spirit and not the spirit from the soul or the soul from the body.

As human exist in this visible universe, his passage from the visible to the invisible universe follows a definite pattern of separation of the components in both direction of communication. That is to say that, there comes a time when the body separates completely from the soul. This happens when human lost his life implying that the body component is no longer in existence but human still exist in this visible universe in a visible but intangible form.

After the loss of existence of the body component, there comes a time when the soul separates completely from the spirits. This happens when human leave this visible universe completely. That is, when human is in the invisible and intangible form implying that the soul component is no longer in existence.

Finally, after the loss of existence of the soul component, human existence in the invisible and intangible form which is the spirit component which is in the invisible universe and it is in reunion with it creator.

This implies that the separation of human components in both directions cannot be re-established but human can re-exist in his three components in this visible world if the creator so desires.

It should be noted that the spirit cannot separate from the soul and the soul cannot separate from the body in a single direction of communication as human exist in this visible universe.

As the final form of existence of human in the invisible universe is the spirit component which is in reunion with the creator, I do believe that the spirit of human is formed as a clone from the creator. Meaning that human in the invisible universe has the same potentials as the creator but, he is send to the visible universe by the creator to understand the universe and manifest his power of creativity and control over it.

CHAPTER THREE

HUMAN AND THE ENVIRONMENT

The phenotypical appearance of human in this physical universe is determined by the genetic constitution of the Deoxyribonucleic Acid (DNA). I believe that this DNA was fashioned in a specific definite pattern by the creator during creation of the body component of human. And it is transmitted from parents during fertilization forming a zygote which becomes a new human body component determined phenotypically by its DNA.

This is why DNA is considered by scientists, as the basic molecule of life. For it produces the body component of human and it structure and function are so complex that it is beyond the understanding of human.

Human existence in this physical universe can be divided into two stages. The early stage which involves the period from fertilization and the end of pregnancy and the late stage which involves the period from delivery still when human leaves this physical universe.

As human comes into existence in this physical universe from conception, he is created to be in a positive state of equilibrium with the environment. This means that, human is not to have any negative influence on the environment and the environment on the other hand is made not to have any negative influence on human.

The positive state of equilibrium can be seen in the early stage of human existence where in a normal pregnancy, the baby (human) has no negative effect on the mother (environment) and the mother doesn't also have a negative effect on the baby.

The late stage of human existence also involves him continuing to be in a state of positive equilibrium with the environment (nature). For human purpose of existence in this physical world is to understand the environment which will in turn help him to understand himself, for he is part of the environment which comprises of every creation (nature).

This state of positive equilibrium can also be seen for example in the relationship between a plant and human where the product of metabolism of a plant such as oxygen, fruits and tubercles are necessary for human development and the product of metabolism of human such as carbon dioxide and components of feces are also necessary for plant development.

As human leaves from his early stage to his late stage of existence through birth and continues developing the body component, he realizes his ability to distinguish the actions that will have a positive or negative effect to the environment. This ability that human realizes as he develops the body component is what is termed "conscience" which in other word is the spirit component.

The environment was made to provide comfort and the basic needs necessary for human development and evolution meaning that, the environment is there to provide a positive effect on human.

But, as human develops with the realization of his abilities and control over the environment since the spirit component is a clone from the creator and has the same potentials as the creator, he turns to look for self-satisfaction (known as sin) and power from the environment. This desire for self-satisfaction leads to a negative effect on the environment which destroys it.

These destructions can be seen in the killing and sufferings of humans by human, the massive killing of other creatures by human to be used as food or construction materials for modernization, destruction of natural habitat of other creatures and pollution of the environment.

The self-satisfactory nature that human develops makes him forget the fact that his existence in this physical universe is fashion by the creator such that, the environment will provide him with the basic needs and effect for his development and evolution respectively, for it has a positive effect on both the genotype and phenotype of human and that each creature has a role to play in this interaction between human and the environment.

So, due to the negative effect that human develop on the environment, the environment also develops a negative effect on human in order to maintain the equilibrium interaction between human and the environment.

This can be seen in diseases that affect human genotype and phenotype which are of infectious, radiation, tumoral and traumatic origins, environmental catastrophes such as earthquakes, tsunamis, hurricanes, flooding, social unrest such as protest and wars. Which all have a negative effect on human development and evolution in this physical universe.

From all these, it can been seen that human is the origin of every negative effect on him and the environment.

Due to the negative effect of the environment on human, there comes a stage where the interaction between human and the environment is disrupted such that both becomes irresponsive to each other in this physical universe (coma) and as time goes by, the disruption can be reestablished or human loses his body component.

CHAPTER FOUR

CREATION AND EVOLUTION

In the beginning, every species of creatures that is found in this physical universe was created by the creator in a process called creation with each having a particular purpose to fulfill.

As time goes by, the interaction between a creature and the environment causes changes in the genotype and phenotype of the creature in a process called evolution which leads to the production of a diversity of that creature.

Due to environmental hazard, there is extinction of some creatures meanwhile there is continues diversification of other creatures. But following the process of creation, there will come a time when the existence of all creatures in this physical universe will come to an end.

These processes of creation and evolution can be represented as a diagram as seen below.

CREATION	EVOLUTION
Beginning ↓ End Involve Existence of Species	Beginning ↓ End Involves Development of a Single Species

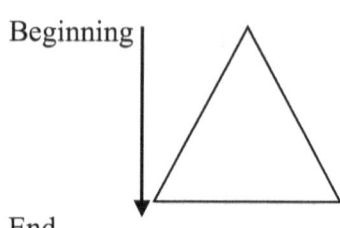

This implies that in this physical universe, there is no more creation for everything was created in the beginning. But there is the process of evolution of creatures forming the diversity of creatures that exist.

Concerning creation, human is the highest creature created by the creator in this physical universe because he exists in the non-physical and the physical universe. The other creatures exist only in this physical universe having as purpose to help human understand creation and most importantly himself which is human purpose of existence in this physical universe.

This can be seen in the fact that human is the only creature that has the ability to carry out thinking, imagination, creativity and communication through speaking. Meanwhile other creatures do not possess these abilities of thinking, imagination, creativity and communication through speaking because they don't have the spirit (conscious) component, but instead have the Soul (subconscious) and the body (unconscious) components.

When human was created, the genotype and phenotype were fashioned such that as human interacts with the environment as time goes by, the effect of the environment on human will cause new actions of the genotype on the phenotype which is responsible for human adaptation and evolution in this physical universe.

These can be seen in the fact that, humanity has a diversity of human race with different phenotype, different languages and cultures, different learning abilities, different geographical areas with different environmental conditions.

Since the purpose of human in this physical universe is to understand creation, the process of evolution of human during the early times after creation was based mainly on the act of procreation. This was to produce a greater number of humans in other to understand all of creation for a single human cannot understand all of creation, but needs the sum total of understandings of humanity.

This is why in the early periods after the creation of human, there were; first and second degree marriages that is to say, marriages between fathers and daughters, mothers and sons, brothers and sisters, nephews and nieces, there was high level of fertility in humans, there was no or low rate of same sex relationship in humans and human lived for a longer period of time. All of which were needed for population expansion.

As the population expanse as time goes by, the environment makes it such that there is a gradual restriction in population evolution.

This can be seen in the present restriction of first and second degree family marriages, occurrence of high level of infertility in humans, increase rate of marriages between humans of same sex, human living for a shorter period of time and a host of other aspect involve in population restriction.

Concerning the aspect of same sex marriage which is increasingly occurring, I believe that, following the evolution of human, human is reproduce to have attraction for same sex or possess characteristic of the opposite sex because the population of human has reached it maximum which is more than sufficient to understand creation. So, this comes in to limit the population expansion for the act of procreation is not human purpose in this physical universe but a means of evolution.

All of these pave the way for human to achieve his purpose in this physical universe which is understanding creation. For as there was a beginning for the existence of creatures, there will be an end in the existence of creatures in this physical universe.

CHAPTER FIVE

RELATIONSHIP BETWEEN THE BRAIN AND THE HEART

The body which is the visible and tangible component of human in this physical universe is determined by the well fashioned genetic composition found in the DNA, forming different structures called body parts which carry out different actions that the body performs.

Amongst the different body parts are the brain and the heart which I believe are the most complex body parts whose functions when altered have a serious negative effect on the entire body. This complexity can be seen with reference to the function they perform, for the brain is a single structure that coordinates the whole body and the heart is a single structure that supplies blood to the whole body.

The brain is the structure in the body that houses all the three components which make up human for his existence in this physical universe. This can be seen in the fact that activities of all structures in the body are coordinated and directed by the brain for it is responsible for the generation of thoughts (spirit component), production of impulses (soul component) which then travels to each structure leading to its action (body component).

It can also be seen in the scientific language that the word conscious, subconscious and unconscious are attributed to processes concerning the brain.

This means that other body parts of human comprises of two components which are the impulses and the action they perform.

Since the brain houses the three components which make up human in this physical universe and it is responsible for coordination of all activities of the body, it doesn't carry out the expression of these three components. So another structure which is the heart through complex connection with the brain is responsible for the expression of these components in this physical universe.

This can be seen in the fact that situations in the brain caused by individual thinking, frightful dreams and environmental conditions such as; events in the family, social relationship and interaction between creatures, will be manifested or expressed by the heart through increase or decrease in blood supply to different structures of the Body if any action is commanded by the brain.

This relationship between the brain and the heart can be seen in our daily lives in which we use phrases such as: I love you from the deep of my heart, he breaks my heart, his heart is as strong as a stone, he has a good heart and his heart is full of sins (self-satisfying nature of human).

All of these phrases refer to the mind located in the brain, for it is the origin of thoughts which leads to the generation of feelings. But since the manifestations of these feelings are primarily express at the level of the heart, which is why the phrases are referred to the heart by human.

But instead, the real phrases should be: I love you from the highest thoughts of my mind, he breaks my mind, his mind is as strong as a stone, he has a good mind, and his mind is full of sins. It should be noted that some of these phrases are just a form of expression for communication.

So, due to the complex communication between the brain and the heart and the function they perform, they are the highest structures in the body for the brain houses the three components of human and the heart manifests these components for human existence in this physical universe.

CHAPTER SIX

SLEEP AND DREAMS

Sleep is a state induced in a human in his existence in this physical universe where by, there is temporal disruption of communication between him and the environment such that the body is irresponsive and carries out no action on the environment. This state of disruption is reverse by the induction of the process of awakening.

During the induction of sleep, there is a transitional period where the body is still responsive from the environment and can also carry out some actions on the environment.

Sleep contains a state called dreaming where the spirit component of human is the dominant component functioning, without commanding the body through the Soul to carry out any action on the environment for the body component is paralyzed.

Since the spirit is the dominant component functioning during dreams and it is the component that interact with the invisible universe which is forever present, that is why some dreams are about things that happened in the pass, others about things that happen in the present, others about things that will happen in the future and others about things that will never happen in this visible universe.

So, dreaming is a state of human in this physical universe where human exist only in the non-physical universe.

During the period of sleep when the spirit dominates which is the main component of human, it carries out repeals and reprogramming of the body component through the intermediary of the soul component.

This can be seen in the fact that, it is during sleep that the fight against aggressions in the body increases, for the strength of the immune system is increased due to reprogramming. Also, wound healing is accelerated and memories are analyzed and stored. This is also the reason why the body component is irresponsive to the environment.

CHAPTER SEVEN

THE MENTALLY OFFSET HUMAN AND THE SOCIETY

The mentally offset human is a state in a normal human being when there is lack of a reversible communication of information between the spirit and the soul.

In a normal human being, the body connects the external environment to the soul and the soul then connects to the spirit. So, information of things coming from the environment is being communicated to the soul and from the soul to the spirit with a reversible pathway.

I do believe that the mentally offset human has a disconnection of the passage of information from the soul to the spirit, so information goes only in one direction; that is from the spirit to the soul then to the body.

This state can be caused in a human through induction of the soul by another human, using electromagnetic brain wave which has a low frequency and travels with the speed of light to attain the human in target, causing hyper stimulation of the soul leading to disconnection of communication of the soul to the spirit.

This state of mental offset can also be caused through natural means, due to the negative effect of the environment on the body (genes, neurobiological structures) which then hyper stimulate the soul leading to disconnection of communication of the soul to the spirit.

Concerning the mentally offset human and human laws, naturally, human is created with human rights in which he is free to do anything at any time and at any place, for he knows the good and bad effect on the environment of any act he performs.

But due to the unnatural acquired characters of human, the society is governed by human laws in other for him to live in sociality and justice which then has as consequents, a limitation to his natural human rights by influencing the action of his spirit.

In the mentally offset human, since the environment does not influence the action of the spirit due to the disconnection of communication from the soul to the spirit, he directly acts by spirit which makes him to have natural human rights and do not respect human laws.

With reference to this, the mentally offset human is classified by the society as being mentally ill (mad human). For example, the mentally offset human walks naked, talks to himself, live in unhygienic areas, feed on inappropriate substances and so on.

Talking about the spoken language of the mentally offset human, it is considered to be inappropriate and useless by the society. But I wish to say that language is something that human learns at any age from when he comes into this physical universe still when he will leave. It can be arranged in any pattern given that a message is passed out which can be understood.

Since the society cannot understand the language of the mentally offset human, it is considered to be useless. But I do believe that his language is understood by another mentally offset human or a normal human with greater dominants of the spirit components.

If I may ask, 'we do have different languages in the universe, why is the society not considering each other's language as being useless and inappropriate?'

I can say that, even some persons in the society with same language pattern do use phrases which are not understood by others but it is understood among themselves.

For example, the politicians; most of the spoken words of the politicians are things that they will not fulfill but they speak them in a metaphorical manner in which the society does not understand. So why are they, not considered mentally ill?

With reference to the confinement of the mentally offset human in the so called mental facilities, I do believe that it is unjust. I believe that since they no longer respect human laws in order to live in sociality with the society, an area should be created for them where they have the freedom to express their natural human rights, interact amongst themselves and are not controlled by the society, but are provided with the necessary needs by the society.

The mentally offset human do not need psycho pharmacotherapy for it has no effect on his state. I do believe that nature at it appropriate time, with the help of a normal human support through psychotherapy and other means, will restore the equilibrium state lost in the mentally offset human.

Concerning health state in the mentally offset human, it is observed that the mentally offset human hardly get sick due to the fact that he hardly contact a disease.

I do believe that this is due to the fact that normally, in a normal human being, the immune system of the body is responsible for the fight against diseases in the body. This immune system is programmed and coordinated by the spirit due to communication of information from the environment through the body by the soul to the spirit, to recognize exogenous or endogenous pathologic substances. When there is disequilibrium within the body in the immune system, there is the appearance of a disease. But in the mentally offset, since they act directly by the spirit and there is no communication of information from the environment to the spirit by the soul, the immune systems is reprogrammed and coordinated by the spirit not to recognize exogenous and endogenous pathologic substances and the presence of these substances in the body have no effect on the body to cause a disease.

This can been observed in the fact that, the mentally offset can drink dirty and contaminated water, eat contaminated food and poisonous fruits, bath in dirty and contaminated water and live in dirty environments without having any health effect meanwhile all of these conditions have a deleterious health effect on the normal human being.

I can conclude that, the mentally offset humans are a special group of people who presence with complete natural human rights and should be allowed by the society to express them in an environment provided for them, without being control but provided for. I believe with time, nature will play it role, associated with our support through psychotherapy, for them to regain their normal human state so that they can again be integrated in the society.

CHAPTER EIGHT

ELECTROMAGNETIC WAVE AND GASEOUS COMMUNICATION BETWEEN INDIVIDUALS

I do believe that there is a system of communication between individuals through an electromagnetic wave (brain wave) and a gas (Nitric Oxide) produced by the brain.

Demonstrations by theoretical arguments show that this electromagnetic wave is the origin of consciousness and that information has been encoded in this wave.

This ability of communication can also help us understand that, as a computer can receive and record the brain wave, so an individual can also receive and integrate this wave and it is being decoded to release the message encoded in it. In this respect, it can be said that the state of consciousness in an individual can be understood by another through concentration.

Some theorists claimed that this means of communication through electromagnetic wave is not possible because the brain wave produced is very weak and cannot travel greater distances. They put up the augments that, brain-waves are too weak to explain telepathy and also that, the electromagnetic field out of the head is far too weak and highly unlikely that any other brain could detect it, and still more unlikely that the other brain could decode the electromagnetic field information that was encoded by your brain.

It was proposed that extremely low-frequency electromagnetic waves may be able to carry telepathic and clairvoyant information. I also do think that this electromagnetic wave has a very low frequency similar to/or lower than that of a radio wave, capable of transmitting information from one individual's brain to another regardless of the distance.

Production of brain wave and the process of communication depend on the fact that; when an individual is in thoughts, action potentials are being generated simultaneously in thousands of neurons due to the passage of ions ($Na+$) across the neuronal membrane.

These action potentials are being propagated in the form of impulses (electric current) through the neurons.

Due to the propagation of these impulses, there is the induction of a magnetic field which is perpendicular to the direction of the electric field created by these impulses. The present of both fields (electrical and magnetic fields) creates an electromagnetic wave that contains the coded information of the thoughts (state of consciousness) called the brain wave.

This wave is then transmitted out of the brain into space where it is being received by the neurons of another individual. It is then decoded to produce the original impulse containing the same information. This can act as an explanation to the phenomenon of telepathy (thought-transference).

The strength of this wave depends on the degree of consciousness and the transmission from one individual to another is time dependent, called the lag time.

This means of communication is observed between students in a classroom who are bench mates. When an unfamiliar question is asked and one of the students thinks of the answer, a few seconds later, the mate also has an answer which is exact to his. It is realized that, this happens in many occasions and with different set of students who are together.

This is associated to latent telepathy which is described as the transfer of information, through psi, with an observable time-lag between transmission and reception.

Also, it is observed that some individuals have a specific conscious mind capable of receiving information from far away distances from those who are dear to them. This can be observed in some cases where; certain individuals of a family, who are far away from home, can feel an effect when their name is mentioned or when an event happens to a family member.

This is associated to intuitive telepathy which is described as the transfer of information, through psi, about the present state of an individual's mind to another individual.

Again, It is observed that a greater percentage of a steady population in a vehicle in motion always get asleep. This sleeping state is always induced by a few individuals and the message is being transferred to others.

Further, it is observed that a mixture of intelligent students and less intelligent students in a group reading which is being characterized by no physical communication, results in the improvement of intelligence in the less intelligent students.

I believe that the actual measuring instrument of brain wave which is an electroencephalogram (EEG) is inaccurate because it measures only the electrical component of the wave. So, I think that a future instrument called an Electromagneto-encephalogram (EMEG) will provide an accurate measurement for brain wave for it will measure both the electrical and the magnetic components of the wave.

With reference to gaseous communication between individuals, during impulse propagation in the neurons of the brain, at the level of the synaptic knob of specific neurons, there is the release of nitric oxide which is a neurotransmitter.

This nitric oxide then diffuses into the blood capillaries surrounding the neurons and it is transported to the lungs where it diffuses into the alveoli and it is expired. The presence of this gas in the outer space influences the thought or state of consciousness of the individual around the area covered by the gas.

This is done through inspiration of the gas which is transported to the brain by the circulatory system where it acts as a neurotransmitter to the neurons specific to it and produces the same thoughts.

Much is not to be said about this means of communication because the quantity of gas produce and the area covered are mild.

Involving the production of click sounds in the brain and flashes of lights in the eyes, the neurons in the brain are connected to each other through specific junctions called synapses. Each synapse contains a potential space called the synaptic cleft that separates the postsynaptic and presynaptic membranes.

During hyperactivity of the brain, there is rapid and intense generation of impulses. The propagation of these impulses leads to intense vibration of the molecular components of the neuronal membrane which leads to the sliding of weakly attached neurons from their junctions on to other neurons. This contact then leads to the production of sparks.

These sparks causes direct stimulation of afferent neurons that leads to the visual field of the occipital lobe. Human is aware of the sparks through impulses transmitted to the retina and being represented as star-like or sharp light.

At the instance of light production, there is also sound production due to the vibration of the molecular components of the neurons and air in the synaptic space. These sounds directly stimulate specific afferent neurons and impulses are being transmitted to the audible field in the temporal lobe.

These explanations can be observed in many situations such as:

- An impact of the head on a hard surface causes star-like light sensation and sharp sound production.
- High concentration on a particular situation for a while is accompanied by flash light sensation with a sharp sound in the brain.
- Epileptic patients having flash light sensation before going into seizure.

Concerning concentration and epilepsies, a highly concentrated individual in thoughts produces a strong wave and a gaseous neurotransmitter. This is then received by neurons of another individual and that same process is induced in him. If the receiver of this wave has a low neuronal activity threshold compared to the producer, the impulses generated will lead to hyperactivity of his neuronal system resulting to the manifestation of epileptic signs. This can explain one aspect of the metaphysical etiology of epilepsy.

I do believe that an individual with low brain activity such as the layman has a very high probability of developing epilepsy compared to an individual with high brain activity such as scientist.

This can suggest that the management of epilepsy can be improved by increasing the individuals' brain activity by subjecting him to concentration needed activities in which his neuronal system will adapt with time.

There are some experimental evidences that neurons do communicate between individuals through electromagnetic wave.

An experiment was conducted in the 1930s to explain this mechanism. In this experiment, the "sender" looks at a series of cards while the "receiver" guesses the symbols.

Also another experiment was conducted which consisted of two individuals at the end of each day having to relax and visualize a mental image or "thought impression" of the events or thoughts the other individual had experienced in the day and then to record those images and thoughts on paper in a diary. The results at the end when comparing one individual' diary to another's, was that "seventy-five percent were found to be correct". From these, the floor was opened for further experiments to be carried out in this field.

CHAPTER NINE

ORIGIN OF GENIUS ABILITY

A genius is the existence of a human in this universe, who has a great imaginative skills which seeks to explain the occurrence of events observe around him. This process of imagination occupies a great part of a genius existence in this universe and it can happen at any point in time in the daily life of a genius with maximum intensity occurring around the mid age of existence.

Due to this imaginative skill in a genius, there is the development of super intelligence, which is the ability of reasoning for question solving which are beyond the scope of an intelligent human.

This process of imagination leads to the birth of ideas which can be transmitted through physical means to become knowledge and this knowledge can be accepted or rejected based on each human level of intelligence.

This process of continues imagination leads to a continuous increase in neuronal connectivity and activities in the neuronal network of a genius brain, which in turn leads to increase imaginative process giving him a positive feedback process of imagination.

The continuous increase in neuronal connectivity and activities in the brain of a genius, leads to side effects in his body and effects on how he interacts with the environment.

Being a genius (having great imaginative skills) is not genetically determined or inherited, but it is an acquired process for a human is not born a genius but becomes a genius.

This can be seen from the fact that the early life history of a genius is below or equal to the normal life process at that period of existence, the biological parents of a genius are never humans with great imaginative skills and the siblings of a genius are never siblings with great imaginative skills.

For a human to become a genius in this physical universe, he has to learn to observe the things happening around him that is to say, observe the things happening in the universe, ask himself questions about the occurrence of things he observe and try to answer the questions by himself using imaginative process and basic knowledge. This will leads to ideas which he can pass out to become knowledge and can then be accepted or rejected by each human base on their level of intelligence.

CHAPTER TEN

THE GENIUS AND THE EDUCATIONAL SYSTEM

Albert Einstein once said that "Education is not the learning of facts, but the training of the mind to think".

The educational system put in place by the society causes a limitation in the development of imaginative skills or great thinking abilities, in the minds of individual humans involve in the educational system.

This limitation is due to the fact that the educational system put in place is based on the learning of pre-existing knowledge also known as "facts" and the individual involve in the system, level of intelligence is assess based on the knowledge acquired through the system.

Due to the assessment of level of intelligence based on the acquisition of knowledge, the individuals involved in the system of education spend most of their time in reading and seeking for pre-existing knowledge. This leads to a limitation in the development of imaginative skills since little time is attributed by the individuals in great imaginative or thinking processes.

This can be seen from the saying by Albert Einstein that "Imagination is more important than knowledge".

In order to under the universe which is the purpose of humanity, each human has a role to play. This role is to produce knowledge about the universe and the sum total of all the knowledge produce by the entire human race (humanity) at a given point in time, will lead to the understanding of the universe and the fulfillment of the purpose of humanity.

With reference to the role that each human has to play in the understanding of the universe, as the individual passes through the different periods of existence in this physical universe, he develops interest in aspects of the universe where he can contribute to produce knowledge.

During the stage of existence when a human has developed interest in aspects of the universe, where he can produce knowledge by developing his imaginative skills, the educational system in which he is involve comes and attribute other areas of little or no interest to him.

This leads to the individual attributing most of his time in these areas of little or no interest, to learn pre-existing knowledge in other to pass through the educational system and this has as consequences a less development of his imaginative skills in his areas of interest of the universe.

Concerning a human with great imaginative skills, there is often conflict between the individual and the educational system he is involved in. Since the educational system is based on the acquisition of pre-existing knowledge, the individual with great imaginative skills (genius) goes beyond pre-existing knowledge and produced his own ideas, which are transmitted to become knowledge and can be accepted or rejected by the humans involve in the educational system.

Due to the fact that the humans involved in the educational system do not have the level of imaginative skills of the genius, there turns to often be a conflict between the genius and the humans in the educational system, which leads to the withdrawer of the genius from the educational system or varying degrees of performance of the genius in the educational system.

From this explanation, it can be seen that the degree of intelligence of a genius can never be measure through any educational means. Also, the educational system should make in such that the areas of interest of each human is been promoted in order to give the individual enough time to develop his imaginative skills to come out with knowledge about the universe.

CHAPTER ELEVEN

GENIUS AND SOCIALIZATION

Socialization is the process of frequent interaction between living things of same species found in the universe. This process of socialization is highly developed in human, since human is at the top of all creatures and the entire humanity has to function as a single unit to attains its purpose of existence which is, understanding the universe.

The process of socialization can be divided into two categories which are; direct and indirect socialization processes. Direct socialization process involves the natural physical interaction between humans and indirect socialization involves interaction between humans through the use of human creations such as machines.

Socialization takes place in varying degrees in different sections of the society constituted by humans. This sections includes; family, friends and community. The process of socialization in all categories and sections is easily acquired in non-genius human.

When it comes to a genius, there is a deficit in the process of socialization in some sections like, the sections of family and friends.

Concerning the family section of socialization, a genius doesn't often have the characteristics of frequent interaction with the family members and this leads to difficulties of the family members in communicating with and understanding a genius.

Concerning the friends section of socialization, a genius has few or no friends. This is because a genius put most of his time in searching to under the universe and a little time in social interaction, which makes him shift away from making friends from the people around him.

A genius is also not too interested in material pleasures of the society which makes him find it difficult in changing the few materials things he possesses and making friends to hang out with to obtain pleasure from materials substances of the society.

When it comes to having intimates relationships, I believe that there is a process of natural barrier between a genius and a human of opposite\same sex. During a state of emotions in a genius, there is the generation of a strong emotional force due to the high brain activity of a genius and this force induces the brain of the opposite\same sex human causing a form of emotional repulsion between them.

This emotional repulsive force combined with the lack of interest of pleasure from societal materials substances by a genius, makes human of opposite/same sex to develop less emotional interest on a genius which has as consequence a state of loneliness in a genius during his momentary free period.

This state of loneliness in a genius can leads to health problems such as depression and the increase attributed time in imaginative processes with high brain activity can leads to schizophrenic behaviours.

It should be noted that, when there is a successful emotional relationship between a genius and a human of opposite/same sex, there is reduction of time attributed to imaginative process and this causes a reduction in imaginative skills with consequent reduction in intelligence of a genius.

CHAPTER TWELVE

THE UNIVERSE

I believe the universe has as origin from a creator also known as a super being. It is made up of two components which are; the non-physical and physical components of the universe which interact with each other for its existence.

The non-physical component of the universe is controlled by the creator meanwhile the physical component of the universe is controlled by a conscious creature in which human belongs.

Everything that exists in the physical component of the universe, originates from the non-physical component of the universe.

The physical component of the universe is govern by the forces of nature put in place by the creator which are in turn influenced by the coarse of time of the makeup of the physical universe.

Time originates from the rotational state of the universe induce by the creator. This state of rotation of the universe is reflected in the rotational state of it makeup such as the galaxies, stars and planets. Time depends on the speed and mass of the physical entity which themselves are determine by the forces of nature.

Due to this, time experience by the makeup of each rotational physical entity of the universe, varies depending on the speed and mass of that entity. This phenomenon of time can only be perceive by a conscious creature in the universe, in which human belongs and the variation of time has a varying effect on each creation of the universe.

Talking about humans, this time variation causes a change in the aging process of human in each rotational physical entity of the universe he finds himself inn. So this causes time distortion in the process of ageing, as human moves from one rotational physical entity to another.

It should be noted that time moves in a single direction in this physical universe and its effect can never be reverse but can be slowed or increase and also that, there is no notion of time in the non-physical universe.

The speed of rotation of the planet earth on its axis is not constant but in an increasing order. This is due to the fact that the environmental condition on earth causes a continuous decrease of the earth mass.

This increase in speed causes accelerated aging of creatures on earth and also, a decrease in the mass of the earth causes an increase pulling effect of the sun on the planet earth causing the earth to go closer to the sun.

This variation in the speed of the earth can explain the reason while in the ancients days, human had a longer life span because the earth was rotating at a lower speed which lead to slower effect of the process of ageing meanwhile in modern age, the earth is rotating at an increasing speed which has as effect an increase in the process of aging in human leading to a shorter life span.

Talking about the probable existence of multiple physical universe and extraterrestrial life by Stephen Hawkings, I do think more of the fact that there is the existence of a single universe with its two components which are the physical and non-physical components.

I go in line with the fact that there is a probable existence of extraterrestrial life (most importantly conscious life) in another planet or galaxy of the universe, for a single conscious life cannot be created only in a single planet of a single galaxy amongst the millions of galaxies and their planets presently estimated by human which make up the universe.

I do believe that the conscious/human life on the planet earth is an experiment by the creator/super being to see the time period it will take, for the entire generation of humanity to understand the part of the universe in which human belongs which is human purpose of existence. In line with this, there is probability that there are other experiments on conscious life by the creator in other galaxies of the universe.

CHAPTER THIRTEEN

TIME TRAVELING

The universe is made up of two components which are the non-physical (invisible) and the physical (visible) universe which interact with each other.

Human exist in both component of the universe due to his constitution into three parts; the Spirit, which acts in the non-physical universe, the soul, which acts in the non-physical and the physical universe and the body, which acts in the physical universe.

Due to the existence of these two components of the universe, human has the ability to time travel in the future and in the past in the non-physical universe and return in the present in the physical universe where he can change the course of time.

I believe that traveling to the future means going ahead the course of time and experiencing events in the non-physical universe that will happen in the physical universe and then returning in the presence to witness those events unfold. With this, human has the ability to change the course of time in this physical universe.

This process of time traveling usually happens during the dreaming stage of sleep and during day time visions. Where a human, using it spirit component, goes into the future in the non-physical universe and experiences an event and then return in the present in the physical universe (wakes up from sleep). He then gives an explanation of the processes that will be involve in the occurrence of that event in this physical universe, which if considered can change the course of that event happening.

For example, if a person dreams of someone having a highway road accident and then wakes up and give an explanatory advice to that individual not to travel. If the advice is considered, it means that the accident will not occur, meaning that the time course has been changed.

But if the individual doesn't consider the advice and the accident occurs exactly as explained, it means that the time course has not been changed and a proof that the dreamer time traveled in the future.

Also, I believe that traveling in the past means going behind the course of time and experiencing events in the non-physical universe which have already occur in the physical universe and then returning in the present.

This means that the time course in this physical universe cannot be changed when we travel in the past because time moves forward and not backward in this physical universe. This is in line with the argument of some philosophers who said that it might be possible to go back in time but it is impossible to actually change the past in any way.

From this process of time traveling in the non-physical universe, we can deduce that not everything that occurs in this physical universe can be proved and that everything that happens in this physical universe happens for a reason because the two components of the universe are in action.

Talking of the possibility of time traveling in the future or past in this physical universe, it means that there have to be the existence of two earth planets in the universe with same creation but different time laps. Meaning that a human exist in pairs: that is one in each earth of the universe.

Due to the time laps between the worlds, a human can travel in the future or the past from one earth to another using a time machine, since both earth are undergoing same processes. So, to me, I believe that the existence of two earths with same content is not logical. Therefore, time traveling in this physical universe can't exist.

CHAPTER FOURTEEN

TIME DISTORTION

I believe that time distortion is gaining or losing time with reference to the interaction of two objects which have different time rates in this physical universe.

The time period of the earth is determined by it complete rotation on its axis giving a day (24hours) and biological processes on earth depend on the time changes as the earth undergoes complete rotation leading to the process of aging as time goes by.

This is in line with the equivalence of biological aging and clock time-keeping mentioned in the twin paradox, where the biological aging is in no way different from the clock time-keeping, meaning that biological aging would be slowed in same manner as a clock.

Considering now the fact that, a human as a biological process leaves the earth to another planet far away from earth which has for example a slower rotational rate of about 10 earth days to complete a single rotation on it axis. Since the biological process depends on the time changes, it means that he will age slower on that planet.

This implies that a 10days spend on that planet will be equivalent to a 100days spend on earth. So when he returns on earth, he will be younger by 100 days than another human on earth who had same earth birth day like him.

From this, it can be seen that human has gained time on earth but he did not travel to the future on earth, because he was in another planet and he did not experience the events that he has met on earth presently before.

I believe that traveling in the future in this physical universe, means going ahead the course of time and experiencing events in this physical universe and then returning to the present to witness those events occur, which is impossible for things occur with time in this physical universe.

Therefore, time machines can be constructed to move human to another planet which has different time rate with reference to the earth so that on return to earth, they can gain or loss time depending on the rate of movement of that planet compared to the earth. This leads to time distortion compared to other humans on earth.

Following Einstein theory of special relativity, the more an object moves faster, the shorter the time it takes. Therefore, if an object can be made to attain the speed of light to move it from one point to another, little or no time will pass by compared to another object that moves at a slower rate. This can be involved in the principle of teleportation.

I believe that an object which can be made to attain the speed of light or higher speed will become invisible and then exist in the non-physical universe. If human can be made to move with the speed of light or higher speed, he will become invisible and can then have a flash of the future or past depending on the time he spend in that state since he will exist in the invisible universe.

This can be seen in the metaphysical disappearance and appearance also known as teleportation of some humans from one spot to another. But this is presently above human understanding to experimentally produce such a state because it goes contrary to the present laws of physics.

CHAPTER FIFTEEN

ORIGIN OF GRAVITY AND WEIGHLESNESS IN SPACE

In the nineteenth century, Isaac Newton put in evidence the existence of gravitational force through observation of a falling apple and then brought out a formula for its calculation, but he could not explain the origin of the gravitational force.

In the twentieth century, Albert Einstein came up with the theory of general relativity which explains the origin of gravitational wave which produces gravitational force. He explains that a moving object in space causes a dimple curvature in space-time know as gravity.

Following the origin of gravitational wave and force, I believe that they are not produce just by the simple movement of an object in space but instead, by the rotational movement of that object on a particular axis in space which greats a gravitational wave in space that induce a gravitational force inside the rotating object.

The strength of this gravitational wave depends on the mass and speed of the rotating object. I believe this is the cause of the differences in gravitational forces between the different rotating objects in the universe.

This rotational nature of an object in the universe causing dimple curvature in space-time depending on the mass and speed of the object, can be known as the "modified theory of general relativity", or the "unified field" since it involves principal substances of nature which are wave, force, mass, speed, time and space.

In the Solar system of the galaxy the Milky Way, where human is found, each planet revolves around the Sun in an orbit and it rotates on a particular axis. It is the rotation of this planet on its axis which creates gravitational wave which then produces the gravitational force which acts inside the planet.

Also, the revolving body around a planet also rotates on an axis which generates its own gravitational wave and force which acts inside the body. The strength of this wave depends on the mass and speed of the planet or body.

In space, there is no gravitational force and the speed of the gravitational wave produce by the rotating objects decreases as it propagates in space with subsequent abolishment. The absence of gravitational force in space causes the objects to be weightless and the equal distribution of gravitational force inside the rotating objects causes it to stay in its axis.

Concerning the weightlessness of Astronauts inside a rocket, I believe that is due to the absence of production of gravitational waves due to the lack of rotation of a rocket in an axis as it moves in space. This leads to the absence of gravitational force inside the rocket which leads to weightlessness.

In reference to this, the principle of rotational movement can be applied on space machines in other to generate gravitational wave in space with subsequent production of gravitational force inside the machine. This will lead to the elimination of weightlessness and all its consequences on the astronauts and a variation in the ageing process of the astronauts depending on the speed of rotation of the space machine.

CHAPTER SIXTEEN

DERICIAN TRIALISM

The world Trialism is used in Christian theology, which is the doctrine that human, is made up of three components which are; the Spirit, the Soul and the Body.

Trialism was introduced in philosophy by John Cottingham as an alternative interpretation of the Cartesian Dualism (Mind-Body Dualism) of Rene Descartes, which states that human is made up of two substances; the Mind and Body, which are distinct and separable, with the Mind being a non-physical substance which holds consciousness.

In Cartesian Trialism by Cottingham, he kept the two substances in Cartesian Dualism and introduced a third substance or attribute called Sensation, which belongs to the union of the Mind and Body.

Going in line with the three attribute (Mind, Sensation, Body) nature of human by Cottingham, I think that the third substance (Sensation), is limited in explaining the processes that takes place between the Mind and the Body. This is because Sensation is based mostly in the perception of the senses and doesn't take into consideration the sub-thinking processes involving memories, emotions and reflexes.

I think that the substance Sensation, in Cartesian Trialism should be substituted with the substance term **"Submind"** in what I called **"Derician Trialism"** which involves; the Mind, Submind and Body. This Submind is equivalent to the Soul component in Christian Trialism and the Subconscious state in Neuroscience.

The Mind was defined by Descartes as a non-physical substance which is involved with the attribute of thought. He identified the mind with consciousness and said that it can exist outside the body.

This is compatible with the Spirit component in Christian theology which is said to be non-physical, existing in the invisible world in communication with the creator, through which everything arises and the main determinant of bodily actions.

This aspects of the Mind goes in line with the thoughts of idealism which states that the mental processes (the Mind/Consciousness) are the origin of physical/material processes (bodily actions) and also with the thoughts of Anaxagoras (480BC) that all things were created by the Mind and held that the Mind held the cosmos together and gave human beings a connection to the cosmos or a pathway way to the divine.

The Mind is also involved with imagination which is the process of higher thinking giving rise to **"high level**

consciousness". This process of imagination is the attribute which makes the Mind to be attributed only to human.

Concerning the aspect of neuroscience, the Mind is comparable to the conscious state of being which neuroscientists are struggling to define in terms of physical processes. But I think that this can't be defined because it's more plausible that mental processes are the origin of physical processes and not physical processes being the origin of mental processes.

The Mind is there for the substance or component involve with the attribute of higher thinking (imagination) which gives rise to high level consciousness. This is equivalent to Access consciousness (A-consciousness) proposed by Ned Block, higher-order consciousness proposed by Gerald Edelman and a combination of self-consciousness and cosmic consciousness proposed by Richard Bucke. This attribute makes the Mind to be attributed to human only and not other animals.

The Submind is the intermediary substance which unites the Mind and the Body and has both non-physical and physical properties. I believe that it is involve with the process of lower thinking including memories, sensations, emotions and reflexes which gives rise to **"low level consciousness"**, defining the non-physical aspect of the Submind. It is also involve with the process of neuronal activities giving rise to impulses which defines the physical aspect of the Submind.

Cottingham proposed Sensation as a third attribute/substance in Cartesian Trialism, but I think that Sensation alone is limited because it doesn't include memories, emotions and reflexes. So, it should be substituted with the attribute/substance which I called Submind which involves the processes of sensations, memories, emotions and reflexes.

The Submind should be considered as the second attribute/substance in Trialism with the Mind and Body being the first and third substances respectively, giving rise to what I term Derician Trialism.

This Submind is compatible with the notion of the Soul component in Christian Trialism which is considered as a link between the Spirit and the Body, the essence of life and separable from the body during death.

Considering the aspect of neuroscience, the Submind is compatible to the Subconscious state of being which deals with the aspects of memory, sensation, emotions and reflexes. I think that the Submind rather than the Mind, is involves with the phenomenon of functionalism in which mental processes are reduced to neuronal activities, which I think just defines the physical aspect of the Submind.

Taking into consideration the different fields of thought (Philosophers, Theologians and Neuroscientists), I can summarize that the Submind is the **"body"** of the Mind and the **"mind"** of the Body. This means that, the Mind and the Submind can act separately from the Body as it happens in some aspects of dreams and the period after death (a Ghost). Also, the Submind and Body can act separately from the Mind as it happens during the processes of reflexes, memories and sensations.

The Submind is therefore the substance involve with the processes of lower thinking including memories, sensations, emotions and reflexes which gives rise to low level consciousness. This consciousness is equivalent to phenomenal consciousness (P-consciousness) proposed by Ned Block, primary consciousness proposed by Gerald Edelman and simple consciousness proposed by Richard Bucke.

Taking into consideration all these explanations about the Submind, it is therefore a substance which exists in humans and other animals. This also goes in line with the "Cambridge Declaration on Consciousness" signed by eminent neuroscientist which states that animals also have consciousness.

I believe that the consciousness of the Submind is transmitted as encoded information in brain waves which are electromagnetic waves that can explain the process of telepathy. The consciousness of the Submind is what can be attributed also to "Artificial Intelligence" and "Philosophical Zombies".

The Body is the third substance in the Trialism nature of human. It is defined as the physical or material substance which exists in the visible world and it is involved with the actions carried out by human which marks the existence of living. The body doesn't think which makes it to be non-conscious in which I attributed the term **"unconscious component"**.

The Body is considered as same substance in philosophy, theology and neuroscience. But there are some idealist philosophers who believe that the materials substances in the universe including the human Body don't exist, for it is an illusion.

But I think that physical substances do exist and there is the non-physical and physical universe which acts together as a single entity with two components, giving rise to the **"dualistic nature of the universe"**.

The Body is therefore a substance which doesn't have consciousness and it is found in both human and other animals.

The three components of human interact in a definite pattern for human existence in this physical world. The Mind interacts with the Submind and the Submind interacts with the Body. I believe that there is no direct interaction of the Mind with the Body.

The Mind can acts separately from the Submind and Body in the invisible world such as in the case of some dreams in which human have no memory of.

I believe that the Submind cannot act separately on its own. Either it is together with the Mind in which it acts as the "body" of the Mind such as in the case of memorable dreams and the period after Body death when the Mind and Submind separate from the Body to form what I consider as a **"Ghost"**. Or it is together with the Body in which it acts as the "mind" of the Body such as in the case of memories, sensations, emotions and reflexes.

The Body cannot exist separately in this physical world because it is non-conscious and will always need the Submind for it to carry out its actions of living.

I believe that there are three stages of existence of human in the universe which are; the physical stage in which the Body is the main actor, the intermediary stage in which the Submind (Soul) is the main actor while the Body exist no more and the non-physical stage in which the Mind (Spirit) is the last actor while the Submind exist no more.

It should be noted that the combination of the Mind, Submind and body gives a Human, the combination of the Mind and Submind while the Body exist no more gives a

Ghost and the Spirit is left alone while the Submind exist no more.

In summary, "Derician Trialism" which comprises of the Mind, Submind and Body, the Mind is the first component which is involve with imagination giving rise to "high level consciousness", the Submind is the second component which is involve with memories, sensations, emotions and reflexes giving rise to "low level consciousness" and the Body is the third component which has no consciousness and is involve with performing actions which marks the state of living.

These components interact in a linear pattern and the three components are found in human meanwhile just two components (Submind and Body) are found in other animals.

CHAPTER SEVENTEEN

PROCREATION EVOLUTION OF LIVING ORGANISMS ON EARTH

I believe that living organisms on the planet earth and the conditions for their procreation and growth were created (designed) by a super being (intelligent designer), and not by the process of spontaneous occurrence. I believe that there are components which makeup each living organism for it growth and evolution on the planet earth.

A Plant is made-up of a component which is; a Body component, comprising of a DNA (genetic makeup) which determines its growth and evolution pattern base on the conditions design by the super being.

An Animal is made-up of two components which are; a Body component, comprising of a DNA (genetic makeup) which determines its growth and evolution pattern base on the conditions design by the super being and a Submind (Soul) component, which coordinates the Body and gives the instinct nature of an animal leading to a state of subconsciousness in an animal.

A Human is made-up of three components which are; a Body component, comprising of a DNA (genetic makeup) which determines it growth and evolution pattern base on the conditions design by the super being, the Submind (Soul) component, which coordinates the body and gives the instinct nature of a human leading to the state of subconsciousness in a human and the Mind (Spirit) component, which controls the Submind and gives the imaginative nature of human leading to a state of consciousness in human.

For growth and evolution of living organisms on the planet earth, the components which makeup living organisms are introduce in a stage manner during the process of procreation. The Body component is introduced in a plant, animal and human during fertilization, which grows and evolves due to the conditions put in place by the super being. The Submind (Soul) component is introduced in animal and human during the beginning of embryogenesis and the Mind (Spirit) is introduced in human during the beginning of foetogenesis.

In human during the period of conception, the Body period takes 7 days, the Body and Submind period take 7 weeks and the Body, Submind and Mind period take 7 months giving a total of 9 months of conception period. This notion of 7 days, 7 weeks and 7 months can be known as **"the Rule of 7s"** in human procreation.

CHAPTER EIGHTEEN

CONSCIOUSNESS

Consciousness is defined as the state of awareness or perception of the self or the external environment. But, I think that consciousness is the ability of producing and acquiring knowledge about the universe.

The origin of consciousness dates back to the beginning of the universe, for I believe that there is an intelligent designer of the universe, who placed particular conditions on a chosen planet (the earth) which are being fine-tuned in order to habour the physical existence of human and other living organisms.

Based on the existence of living organisms on earth, I believe that consciousness can be classified into two types which are; high level consciousness and low level consciousness.

High level consciousness is found only in human due to the Mind (Conscious) component of human meanwhile low level consciousness is found in human and other animals due to the Submind (Subconscious) component which they possess.

Concerning plants as living organism, I do think that they do not possess consciousness because they are made up of a fashion body which has the conditions necessary for their evolution, for they have neither a Mind nor a Submind component.

High level consciousness in human is produced from the Conscious (Mind) component of human and the Conscious component of human is produce as a clone from the intelligent designer (Super being) who is the origin of universal consciousness.

I believe that the physical existence of human in this universe is an experiment by the super being to determine the time period the entire human generations will use for complete understanding of the universe.

The conditions guiding the chosen planet are fashion to be in a state of equilibrium with human in order to ease human processes in the understanding of the universe. But the negative actions of human on the environment have led to the difficulties encountered in the process of comprehension of the universe.

Concerning the conscious nature of the universe, I believe that the universe produces consciousness for it contains the non-physical intelligent designer and the non-physical Conscious component of human.

I believe that the Conscious component of human is a clone (self-replication) from the intelligent designer which is introduced as a non-activated component during the foetal period of conception, for the physical existence of human. With the expulsion of human in the universe during birth, the Conscious component is activated in order to achieve its purpose of knowledge production and acquisition of the universe.

The conscious nature of the universe makes it possible for astrological reading of the living process of human in this physical universe. This is because the universe contains the non-physical Conscious component of human which interacts with the universal bodies and is also present before and after human existence in this physical universe.

CHAPTER NINETEEN

THE PURPOSE OF HUMAN EXISTENCE

The question on the purpose of human existence is to seek to explain why did the creator create human being and in other words; why do human exist?

According to Abrahamic religion, they believe that the purpose of human existence is to worship the creator (God) and to carryout procreation, as revealed by the prophets and written in the holy books.

According to the layperson, the purpose of human existence is to obtain the basic necessities of life such as; dressing, feeding and shelter.

According to some philosophers, they believe that human do not have a purpose of existence, for the origin of human was spontaneous and not created by a creator. Others believe that if human evolved from lower animals and other animals don't seek to know why they exist, therefore human should not seek to know if human has a purpose of existence.

According to humans practicing spirituality, they believe that the purpose of human is to attain a state of divinity (enlightenment)

To me, I believe that the purpose of human existence is to understand the universe. The creator of human gave human a physical life so that human can seek to understand who human is and why other things exist. Human understanding who human is will make human to realize the abilities that human has which are similar to that of the creator for, the creator created human in his own image (the spirit component of human is a clone from the creator) and in the end, human will know what the creator is. So, the goal of humanity is to understand the universe and each human has as purpose to produce knowledge which will contribute in the sum total of the knowledge about the universe. This acquisition of knowledge about the universe led to the state of enlightenment.

I believe that doing good to others and not harming others are the moral values that we have to practice in our earthly life, for us to live a happy earthly life due to the law of karma and for us to be qualified to exist after earthly existence. So, I don't believe that our moral values and worshipping the creator (God) are our purpose of existence

CHAPTER TWENTY

HUMAN EXISTENCE AFTER DEATH

For human to exist and live in this visible universe, he is made up of three components by the creator which are; the Spirit (Mind), the Soul (Submind) and the Body components.

During the state of death when the body separates from the soul and spirit, to exist no more by returning to dust, there is the end of life in this visible universe for the body components determine life. But, there is continues existence of human in the visible and invisible universe with the spirit and soul components which is an intermediate or transitory stage of existence between the visible and invisible universe.

In the intermediate stage of existence of human, there comes a time when the soul separates from the spirit component, to exist no more by disappearing, which then marks the stage of existence of human in the invisible universe with the spirit component.

From this explanation, it can be seen that there is no life after death, for life is determine by the body component which no longer exist, but there is existence of human after death due to the existence of the spirit component.

I believe that in the invisible universe where the spirit component of human exist after the non-existence of the body and the soul components, not all the spirit components of humans who lived in this visible universe exist.

At the end of the stage of existence of the spirit and soul components of human, when the soul separates from the spirit component for human to exist in the invisible universe with the spirit component, the spirit of some humans is absorbed by the creator who is a spirit, for I believe that the spirit of human is created as a clone from the creator. Meanwhile, the spirits of other humans is left to exist in the invisible universe together with the creator.

The factors which determine which spirit of human is to be absorbed by the creator or left to exist with the creator, depends on how human lived his life in this visible universe.

I believe that human whose life in this visible universe had little or no negative effect on other humans will have their spirit left by the creator meanwhile humans whose life in this visible universe had much negative effect on other humans will have their spirits absorbed by the creator.

Therefore, I believe that the notion of heaven (place of eternal happiness of the spirit) and hell (place of eternal suffering of the spirit) after death doesn't exist but there is the eternal existence of the spirit components of some humans in the invisible universe.

I believe that heaven and hell exist as human lives in this visible universe and not as human exist in the invisible universe after death. For, I think that heaven is the happy moments meanwhile hell is the suffering moments as human lives in this visible universe.

ANNEX

KEY SENTENCES

In this life, you have to believe in yourself and don't let people to be the **main** determinate of your believes.

The main purpose of human on earth is to generate ideas, which can be pass out as knowledge for the understanding of the universe.

The acquired self-satisfying nature of human on earth, is the origin of human sufferings.

Human life on earth will be beautiful, if human respect nature.

Observations of things happening, show us the way in exploiting our Minds to come out with new ideas and discoveries

Creativity is the ability to generate ideas which are applied for production meanwhile intelligence is the ability to think for question solving. Creativity is englobed in intelligence.

In life, humans can differentiate good from bad. So every act we perform, we are conscious of it.

Procreation is an act of evolution of human on earth and not the purpose of human on earth.

We have to think before speaking to someone, for each sentence we say has a positive or negative impact base on his interpretation.

Good means having a positive effect on nature and bad means having a negative effect on nature.

For you to become a great scientist, you have to be a great philosopher. For science originates from philosophy.

Human life on earth is an experiment by the creator, to see how long the human generation will use, to understand the universe.

Human spirit is a clone from the creator, giving him the ability of the creator which he can realize by understanding the universe.

Human suffering on earth can be abolish, if the politician and the rich, change their mindset by respecting human life.

The interaction between human and the environment, is in a reversible state in order to maintain equilibrium of existence.

The beginning of human existence in this universe is the beginning toward the end of life and not the end of existence.

For human existence in this physical universe, he is made-up of three components in what is known as; The trialism Nature of Human.

The cosmos is conscious due to the invisible component of the cosmos, which contains the creator and conscious component of human.

The conscious nature of the cosmos (Universe), makes astrological reading of human possible.

A genius is a human with great imaginative skills. One is not born a genius but one becomes a genius.

The creator is made-up of cosmic energy and can take any form.

www.ingramcontent.com/pod-product-compliance
Lightning Source LLC
Chambersburg PA
CBHW030444220526
45464CB00006B/2407